Fruit Book

Jessica Lee Anderson

Paperback ISBN: 979-8-9899560-9-8

Photo credits—Front cover: A: Billion Photos, B: Darshini_K27, C: Kativ, Aiselin82; Back cover: Komar Gallery; Cover page: Elena Pooto; Copyright page: Anna Kucherova (Papaya); Dedication page: JPC_PROD, p. 4: A: Aniru Khattirat (Annona), Billion Photos, Anna Kucherova, diogoprr; p. 5: B: vm2002 (Breadfruit), Billion Photos, igorartmd; p.6: C: Ekaterina79 (Cranberries), The Branding Co, Billion Photos, ROMIXIMAGE; p. 7: D: qqdigitalart (Durian), Anna Kucherova, Ugreen, olgysha2008; p. 8: E: Avi Rozen (Etrog Citrus), Ratree Fuang, Adap88xx, bdspnimage; p. 9: F: Hyung Jung (Five-flavor Berries), alex-skp, Picture Partners, Ribeiro Antonio; p. 10: G: Mercrades (Granadilla), Anna Kucherova, Ancelin, Buntysmum; p. 11: H: Ozgurkeser (Hackberries), LPETTET, Ariyani Design, Charles Wallertz; p. 12: I: Mabelin Santos (Icaco), MauMyHaT, Author, Paralaxis; p. 13: J: Brasil 2 (Jatoba), PisitPong-14, eyeblink, Adrianeciurea69; p. 14: K: Ajay Kampani (Karonda), kefkenadasi, Anna Kucherova, KADImages; p. 15: L: EMFA16 (Loganberries), Art Studio Images, Billion Photos, Maxsol7; p. 16: M: Kativ (Marionberries), smuay, Phive, Billion Photos; p. 17: N: Chorboon_Photo (Nutmeg Fruit), Ania Lamboiu, Billion Photos, masa44; p. 18: O: slpu9945 (Oil Palm Fruits), GomezDavid, Inna Tarasenko, Danny Smythe; p. 19: P: LisaaMC (Pomegranate), Anna Kucherova; p. 20: Q: mollypix(Quandong), Denny Thurston Photography, JitkaUnv, Arnaldo Robert; p. 21: R: BD Photos (Red Bananas), Anna Kucherova, Billion Photos, thunyakon; p. 22: S: Knaqiyyah (Snakefruit), gresei, Billion Photos, chengyuzheng; p. 23: T: v777999 (Tamarillo), cgdeaw, Anna Kucherova; p. 24: U: Igaguri_1 (Umeboshi), svera, chengyuzheng, Marco Tulio; p. 25: V: fotoedu (Valencia Oranges), Olga Korica, bdspnimage, ValentynVolkov; p. 26: W: insjoy (Wax Apples), canghai76, harneshkp, Billion Photos; p. 27: X: UliU (Xinomavro Grapes), Oscar Yoshinori Toyofuku, Claudio Valdes, Xarhini; p. 28: Y: Yanjun139655 (Yangmei), omega77, Anna Kucherova, manbo-photo; p. 29: Z: Alasdair James (Zante Currants), Margouillat Photos, Nahhan, Andreas Steidlinger; p. 30: Billion Photos, chengyuzheng, Photography Firm, qqdigitalart, p. 31: Michael Anderson

This Book Belongs to:

_____

A is for . . .

Apricots

Açaí (pronounced ah-sigh-EE) Berries

Apples

A a

B is for . . .

Bananas

Blueberries

Blackberries

B b

C is for . . .

Coconut

Cherries

Cantaloupe

C c

 **is for . . .**

Dragon Fruit

Dewberries

Dates

D d

E is for . . .

**Emblica
(also known as
Indian Gooseberry)**

Elderberries

**Elephant Apples
(also known as Chalta Fruit)**

E e

F is for . . .

Figs

Finger Limes

F f

Feijoas (pronounced fay-JO-ahs, also known as Pineapple Guava)

G is for . . .

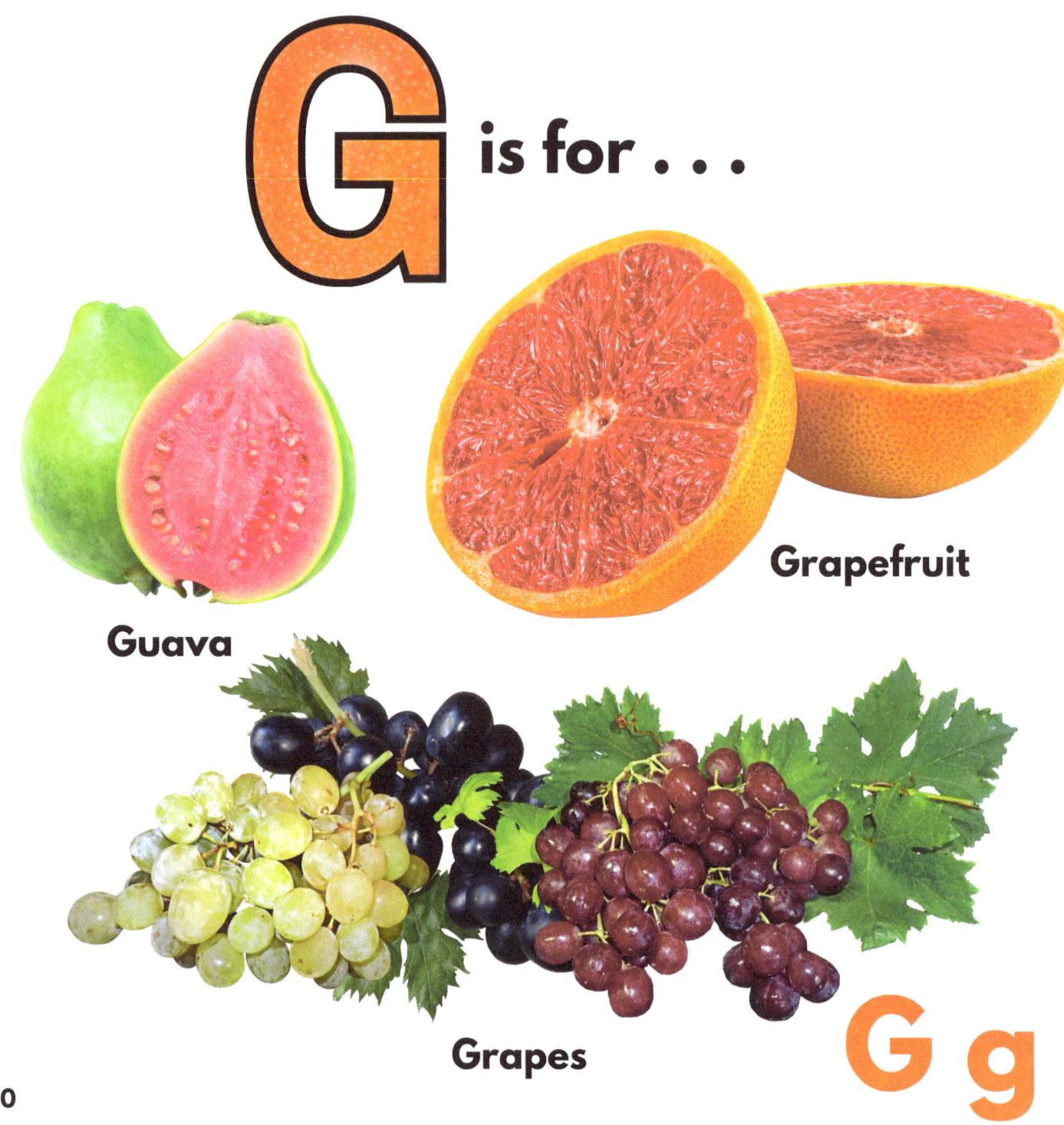

Guava

Grapefruit

Grapes

G g

 **is for . . .**

Horned Melons

Honeydew Melon

H h

Huckleberries

11

I is for . . .

Incaberries (also known as Golden Berries or Uvillas)

Ice Cream Beans

Ita Palm Fruit (also known as Buriti Fruit)

Ii

 **is for . . .**

Jujubes (pronounced joo-joobs)

Jackfruit

Juniper Berries

J j

K is for . . .

Kumquats

Kiwifruit

Key Limes

K k

L is for . . .

Lemons

Limes

Lingonberries

L l

M is for . . .

Mangosteens

Mandarin Oranges

Mangoes

M m

16

N is for . . .

Nectarines

Nectacots

Nashi Pears

N n

 is for . . .

Oroblanco Citrus Fruit

Olallieberries

Oranges

O o

P is for . . .

Peaches

Pears

Plums

P p

19

Q is for . . .

Quinces

Queen Anne Cherries

Quenepa (pronounced ke-nepa, also known as Spanish Limes)

Q q

R is for . . .

Raspberries

Redcurrants

R r

Rambutans
(pronounced ram-boo-tanz)

21

 is for . . .

Strawberries

Starfruits

**Sharon Fruits
(also known as
Non-Astringent Persimmons)**

S s

T

is for . . .

Tamarind
(pronounced ta-muh-rihnd)

Tangerines

Tangelo

T t

23

U is for . . .

**Ugni Berries
(also known as
Murta Fruit)**

Ugli Fruit

Umbú Fruit (Brazil Plums)

U u

V is for . . .

Velvet Apples

Velvet Pink Bananas

Vanilla Fruit

V v

W is for . . .

**Wolfberries
(also known as
Goji Berries)**

White Mulberries

Watermelon

W w

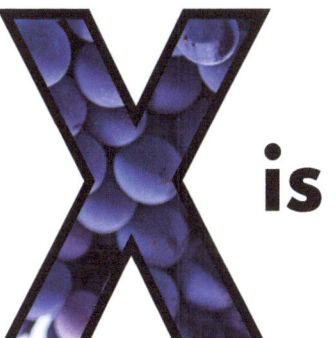 **is for . . .**

Xarel-lo (pronounced zuh-rehl-loh) Grapes

Ximenia (pronounced zuh-MEE-nee-uh) Fruit

Xoconostle (pronounced choko-nose-leh) Cactus Fruit

27

Y is for . . .

Yellow Watermelon

Yellow Passion Fruit

Yuzu (pronounced yoo-zoo) Fruit

Y y

Z

is for . . .

Zebra Melons

Zhe Fruit (also known as Mandarin Melon Berry)

Z z

Zwetschge (pronounced zvetsh-guh), also known as Blue Plums)

5 Fruit Facts:

 Fruits develop from the flowers of plants, and fruits also have seeds.

2 Tomatoes are considered fruits, and they are popular all over the world!

3 Avocado, cucumber, squash, and eggplant are also considered fruits.

4 Not all oranges are actually the color orange! Some may be yellow or even red like blood oranges.

 Some fruits are sour like a lemon, and some are sweet like pears, while others are stinky like durian.

Jessica Lee Anderson is an award-winning author of over 50 books for young readers. Jessica lives near Austin, Texas with her daughter, Ava, and husband, Michael. Fruit is one of Jessica's favorite treats. You can learn more about Jessica by visiting www.jessicaleeanderson.com.

Check out these other titles:

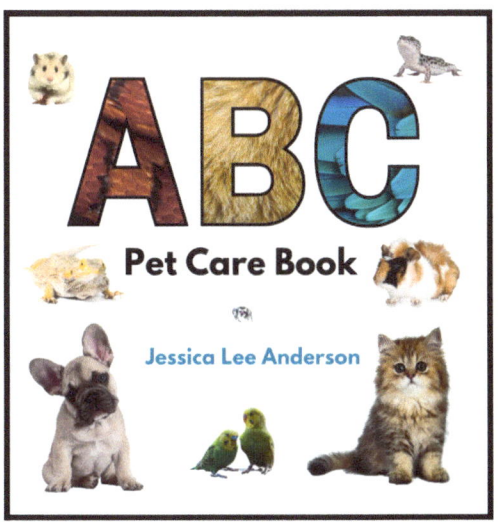

ABC Pet Care Book

Jessica Lee Anderson

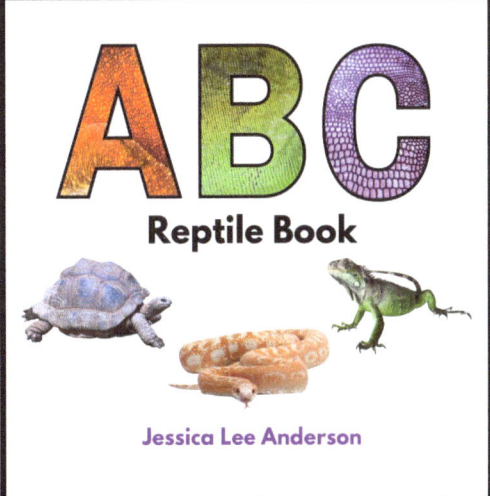

ABC Reptile Book

Jessica Lee Anderson

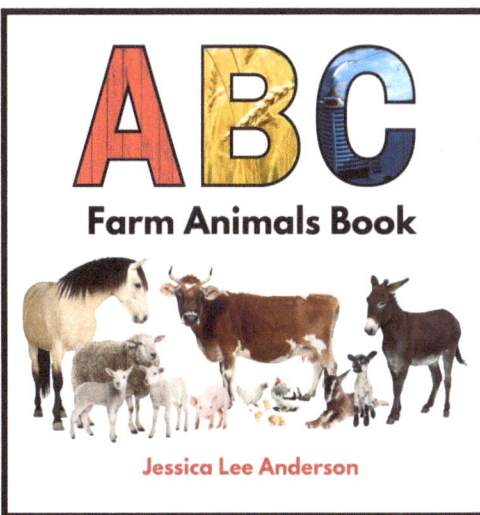

ABC Farm Animals Book

Jessica Lee Anderson